A first guide to
POND LIFE

SIMON PERRY

Illustrated by
Cecilia Fitzsimons

HODDER AND STOUGHTON
LONDON SYDNEY AUCKLAND

For Elizabeth

British Library Cataloguing in Publication Data

A catalogue record for this book is
available from the British Library

ISBN 0 340 56597 7

First published 1993

The author's and publisher's thanks are due to the
following for permission to reproduce photographs:
(t) = top
(b) = bottom
Heather Angel 9, 11, 17, 19, 24(t), 26, 27; Aquila cover, 1,
14(t), 14(b), 16, 18; Ardea London 8, 13, Maurice Nimmo 10;
Oxford Scientific Films 12, 20, 21, 22, 24(b), 25.

Published by Hodder and Stoughton Children's Books,
a division of Hodder and Stoughton Ltd,
Mill Road, Dunton Green, Sevenoaks, Kent TN13 2YA

Typeset by Litho Link Ltd, Welshpool, Powys, Wales
Designed by Trevor Spooner
Printed in Hong Kong by Colorcraft Ltd

Contents

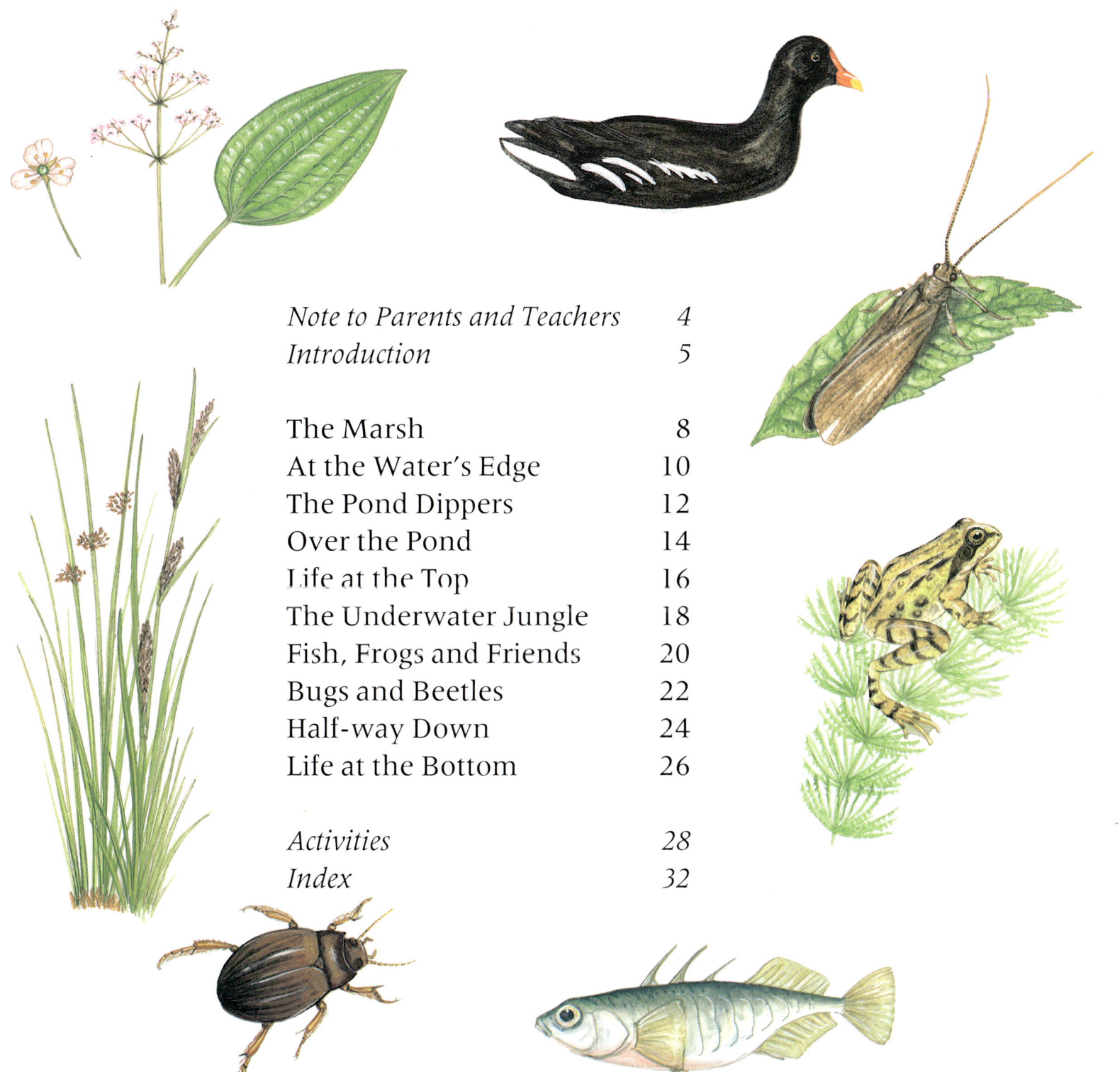

Note to Parents and Teachers

Children of all ages enjoy messing about in water and most of us go fishing, swimming or boating at some stage in our lives. Children are also naturally curious about the strange creatures that live in water: a fascinating wetland world so different from our own. Wetlands come in many different sizes and forms. There are rivers and streams, marshes and bogs, man-made canals and the still, open waters of lakes and reservoirs. But a pond is the most accessible wetland and the one with which children are already most familiar.

Many of our great wetlands have been damaged or destroyed by drainage or pollution, and ponds in the countryside have been filled in. Recently many ponds have been constructed in gardens and school grounds, helping to safeguard something of the rich diversity of aquatic life.

This book views the pond as a habitat, home to a community of plants and animals living together and dependent on each other. While it can be used to identify common pond plants and animals, it will also help children to understand how a pond works. Species illustrated and described are generally common and widespread, including plants likely to be found in newer, man-made ponds. Children may come across similar species which are not illustrated but, with the aid of this book, they should be able to assign them to their correct place and role within the structure of the pond.

A pond is small and fragile and, without great care, there is a danger that a study of one may itself be damaging. It is important that children learn care and consideration for living things, and restrict dipping and trampling. Such common sense is also important for their safety. Conservation and safety are stressed in the Introduction and throughout the book.

A well-maintained pond is something to treasure, but small ponds may turn out to be a temporary feature. Neglected, they fill up with silt, aggressive plants and all sorts of man-made rubbish. Encourage children to use this book as a starting point for their watery adventures, undertaking the activities to understand more. With knowledge and concern, they, and we, can learn to look after our ponds, which are such an important part of our wildlife heritage.

Introduction

What is a pond?

A pond is usually a small area of water. It is not deep like a lake, and its water does not move like a river. However, many of the plants and animals which live in lakes and rivers can also live in ponds.

Older ponds are lined with clay to make them waterproof. Newer ponds may have a concrete or plastic bottom. A pond is shallow at the edge but may be deeper in the middle. It is the habitat, or home, for many types of plants and animals.

Plants in a pond

Marsh plants live in the damp soil at the edge of a pond. Swamp plants grow in deeper water, emerging into the air. Further out you may see plants with floating leaves. Some have roots which dangle in the water. Others have roots anchored on the pond bottom. Below the surface there is a jungle of water-weeds. Many of their leaves are fine and hair-like so that they are not torn by the water.

Tiny plants called algae float in the water. Some algae join together to form long strands, which sometimes make a pond look green.

Animal habitats

Like plants, pond animals live in different places, or habitats. All animals need a gas called oxygen, but there is very little oxygen in water. Pond animals collect air bubbles from the top or have leaf-like gills to get more oxygen from the water.

Moving through water can be difficult because it is thicker than air. Many water creatures are smooth and streamlined. Some have legs shaped like paddles.

Who eats whom?

All green plants can make their own food from chemicals in the water and in the air, using energy from the sun. Herbivores, such as pond snails, eat plants. Carnivores, such as dragonfly larvae, eat other animals. The largest animals are at the top of this food chain, feeding on many of the smaller creatures in the pond.

Many pond animals, such as bloodworms and water hog lice, eat pieces of dead leaves.

Tiny bacteria and fungi also take the goodness from dead plants and animals, causing them to rot or decay. Together these animals and microscopic creatures recycle important chemicals to the water, to be used again by the plants.

There are many food chains in a pond. They are connected together to form a web. *Some* of the animals in the web are shown here connected by *some* of the chains.

A pond food-web

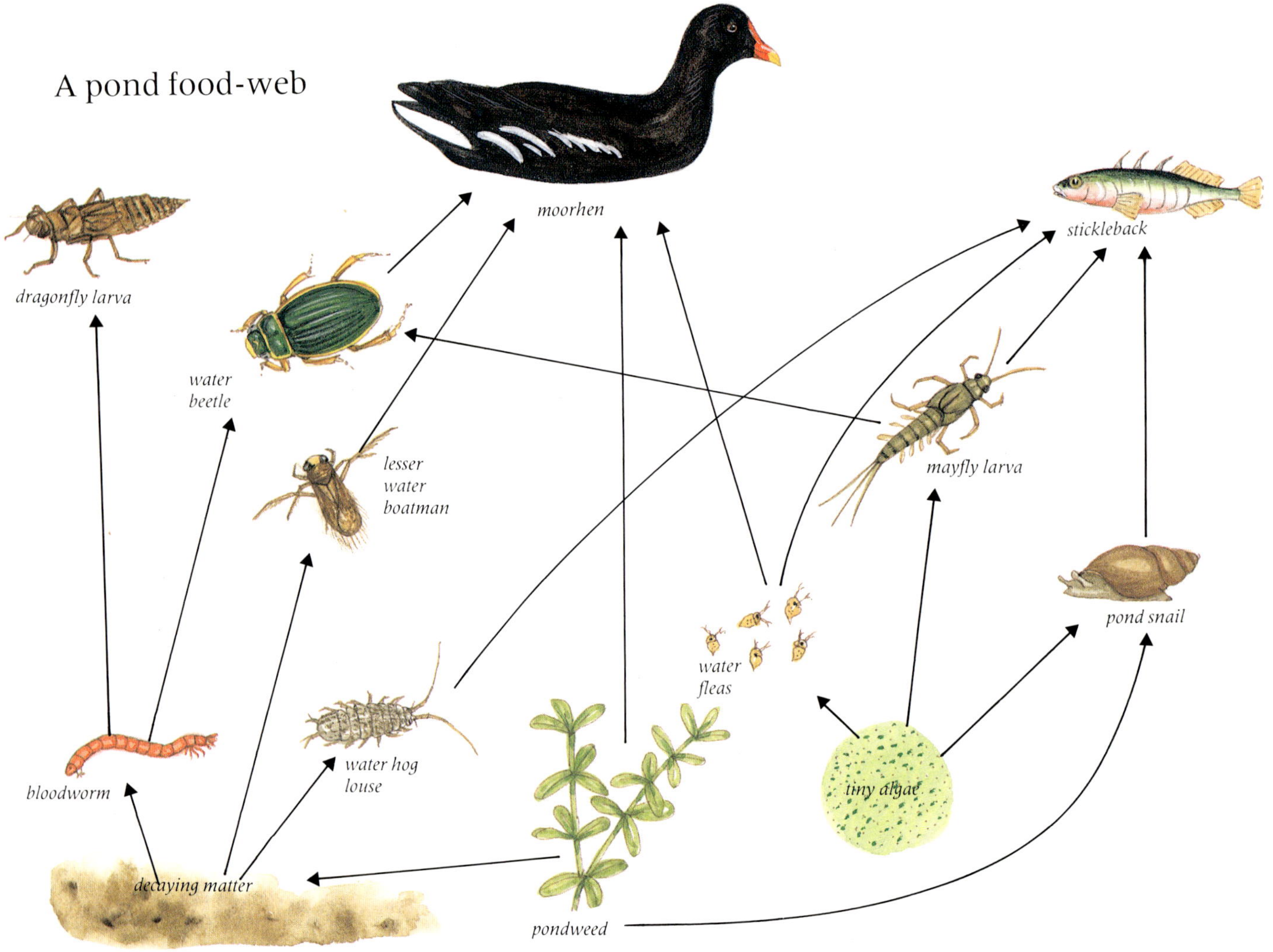

moorhen

stickleback

dragonfly larva

water beetle

mayfly larva

lesser water boatman

pond snail

water fleas

bloodworm

water hog louse

tiny algae

decaying matter

pondweed

Where and how to look

Most ponds are fairly new and nearly all were made deliberately at one time. Older ponds, in the centre of villages, were once filling stations for horses and cattle. Millponds were dug near mills worked by water-wheels. Some ponds were made to provide fish or ducks for food. Many ponds were made by enlarging smaller natural pools, or by damming streams.

This book will help you to identify common pond plants and animals. The first few pages describe the wildlife to be found at the edge of a pond and then its visitors. Next you can read about life on the surface of the water, amongst the water-weeds and in the open. Finally you can find out about the creatures at the bottom of a pond.

A year in the life of a pond

WINTER

SPRING

AUTUMN

SUMMER

Neighbourhood wetlands

Ponds are important for wildlife. Many of the largest wetland areas have been drained or damaged by pollution. Many ponds in the countryside have already been filled in. The ponds that remain, and the new ones we can make, provide a home for many fascinating plants and animals.

Ponds can still be found in parks, school grounds and in gardens and these are the easy ones to study. Take care not to trample the plants, to stir up the mud or to leave any litter. Make sure that any animals you look at are carefully put back into the water before you leave.

REMEMBER the pond-watching safety code:
1. **Always** go with an adult when visiting a pond.
2. If the pond is not at home or in your school grounds, make sure that you have permission to visit.
3. Always stay at the side of the pond. Do not dip where the banks are steep. Always use a stick to test the depth.
4. If you have any cuts or scratches on your hands, cover them with waterproof plasters. You could also wear rubber gloves.
5. While dipping do not eat, drink or put fingers or equipment into your mouth. Wash your hands with soap and tap water as soon as you can after dipping.

Ponds are exciting places to explore, but you must always take care. Follow these simple rules and you will be safe and still have fun.

The Marsh

Plants which grow at the edge of water like damp soil, but they do not like being covered with water all the time. This area is often called the marsh.

▼   Purple loosestrife provides a bright splash of colour in a dense clump, in the marshy land beside this pond. Each spike has short leaves without stalks, and flowers with six petals.

In front are one or two plants of meadowsweet, which may also form dense clumps. Each stem ends in an umbrella-like flower head with tiny creamy flowers. Although the flowers are sweet-smelling, the leaves are not.

Alders and pussy willows don't mind growing in very wet soil. The alder can become a tall tree. Both have catkins which flower in early spring. Willow catkins make a lot of sugary nectar to attract insects. The seeds which are produced have silky hairs and drift in the wind across the pond. The tiny seeds of the alder can float on the water, but in the autumn many are eaten from the cone-like fruits by flocks of small birds.

▶

male

female

alder

alder

pussy willow

male

female

pussy willow

This beautiful waterside buttercup ▼ is the kingcup, or marsh marigold. Its large yellow flowers, seen in early spring, brighten up the edge of the pond. It has large heart-shaped leaves. It is often planted beside new ponds.

soft rush

sedge

rush

sedge

▲ Rushes and sedges are grass-like plants that grow in tufts at the edge of water. Sedges have flat, sharp leaves and grass-like flowers, but the stems are three-sided instead of round. Rushes have narrow rounded leaves with sharp tips. The stem holding the flower is soft or pithy inside.

great willowherb

brooklime

▲ The great willowherb has very hairy stems and leaves and produces bright purple pink flowers. When the seed pods open, masses of fluffy seeds are carried away in the wind.

Brooklime is related to the blue speedwells which grow in garden lawns. It trails across the mud at the edge of a pond, producing new roots and spreading quickly.

Before the arrival of electric lights many country people made cheap candles from rushes. The rushes were cut in autumn, when they were still green. The outer skin was peeled away from the inner pith, which was then soaked in fat. These rush lights burned quite slowly.

At the Water's Edge

Away from the marshy edge of a pond is an area sometimes called the swamp. Most of the plants have roots in the bottom, grow through the water and emerge into the air. They may grow so closely together that it is difficult to see the water.

reeds

bur-reed

▲ Reeds are the tallest plants growing at the water's edge, sometimes reaching over three metres in height. In early August they have purple grass-like flower heads. During the winter the leaves shrivel up while the stems remain as hard canes.

Bur-reeds are smaller, growing to about one and a half metres, with thick three-cornered leaves. The unusual flowers of the bur-reed are like tiny balls, appearing from June to August. These flowers form a spiky round fruit.

▲ One of the tallest plants is the reed-mace, or bulrush, growing up to two metres high. It has large flat leaves and sausage-like flowers at the top of round stems. The male flowers, which produce the pollen, are in the tassel above. In the autumn masses of tiny seeds are released from the tassel, often pulled out by small birds.

▲ The broad spoon-like leaves of the water plantain are easy to recognise. The white flowers each have three petals and are held above the leaves on a taller stem.

▼ Watercress is a type of water cabbage, with many of its leaves growing under water. The lower leaves stay green in the winter. Like many other members of the cabbage family, watercress can be eaten, but it is easy to mistake it for other plants. **Never taste any plants growing by water.** Watercress grown for eating is the same as the wild plant, but it is grown in special watercress beds, in fresh running water.

▲ The yellow iris, or yellow flag, is a large plant which can grow in the marsh or swamp areas. Look for its yellow flowers from May until August, although the broad flat leaves are also easy to spot.

Pond facts

Some reed beds are cut and used for thatching. The tall stems of reed and reed-mace can trap the mud and in time this makes the water shallower. Plants from the marsh will now move in and a small pond will begin to dry out.

11

The Pond Dippers

Many animals visit ponds to drink. House martins collect mud from the shallows to help with their nest-building. Beside undisturbed ponds look in the mud for the tracks of secretive mammals like deer.

▼ What a wonderful sight! The beautiful kingfisher catching small fish, such as the stickleback, with its dagger-like bill. The kingfisher will feed in small ponds, but in towns it is most likely to be seen as a brilliant blue and red flash, darting low along a river or canal. In a cold winter many kingfishers will not be able to catch fish. Some will die or move to other places to hunt for food.

The heron is a large grey bird with ▶ very long wading legs. It stands still in the water, waiting, then with a fast stab of its large bill catches a fish or frog. Herons will visit garden ponds, especially those with goldfish.

▼ The mallard is the most common duck. The pond on the village green or in the town park is likely to have at least one pair. The male is brightly coloured, with a green head, while the female is dull, camouflaged as she sits on a nest at the water's edge. Mallards eat a variety of food, dipping their heads underwater rather than diving. Many are well fed by children with bread and seeds.

▲ Even in town centres, moorhens may be seen on or beside larger ponds. As you go near, they may hide amongst the taller plants, and this is where they nest. Look for the bright red and yellow bill and the constantly moving white tail. Moorhens feed on plants as well as snails, worms and insects.

▼ Shrews have long sensitive noses and tiny eyes, and are always on the move. The handsome water shrew has a slaty-black back with a fringe of hairs on its tail and feet to help it to swim. It makes burrows with holes above and below the water at the edge of a pond.

Pond facts

Water shrews swim along the surface, then dive to the bottom, with air trapped in their fur to keep them dry. They can dive one metre down but can only stay under for about four seconds. They catch shrimps and caddisfly larvae, but this takes a lot of effort so they also catch animals on land.

Over the Pond

Many insects fly round the edge, or over a pond. You may disturb some as they rest on the marsh plants. Colourful dragonflies and damselflies are active on calm, sunny summer days.

▼ During a night in May, June or July a dragonfly larva climbs up on to a tall plant and splits its skin. From this larval skin the four-spotted chaser is emerging. Dark marks form on each wing, which give it this name. Near boggy ponds it darts quickly backwards and forwards, catching insects in the air.

▲ Large numbers of alderflies can sometimes be found in early summer. They fly poorly at night and crawl amongst the marsh and swamp plants during the day. Their dark brown wings are folded like a roof.

▼ These delicate damselflies are related to dragonflies. They skim low over the water to catch small insects. They often settle on water lilies and other floating plants. The male common blue damselfly has a blue and black body while the female is greenish. The male holds on to the female while she crawls underwater to lay eggs for up to an hour.

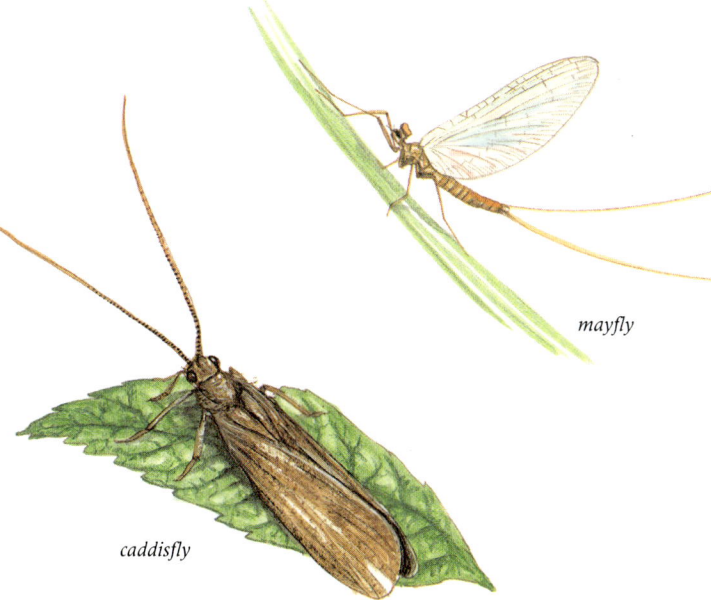
mayfly

female

male

caddisfly

▼ The mosquito, or gnat, and the midge are both types of fly with one pair of wings. Male mosquitoes have very bushy antennae. They feed on the nectar made by flowers. The female mosquito sucks blood from birds and mammals.

▲ The adult caddisfly looks like a dull moth with hairy wings. It has long straight antennae and spurs on its legs. This is the great red sedge. It will only live for a few weeks.

Adult mayflies have two or three long tails and larger front wings. They cannot eat and after they emerge from the water in May and June they lay eggs and die. Many are eaten by fish, birds and bats.

Pond facts

Unlike all other insects, mayflies go through a stage called the sub-adult, or dun. The skin of the larva splits and the dun flies to a stone or plant. Soon the skin splits again, but this time the shining adult, or spinner, comes out.

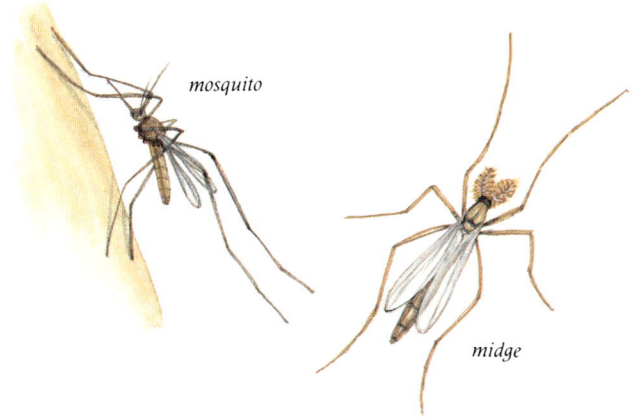
mosquito

midge

15

Life at the Top

In deeper water you will see plants with leaves floating on the surface. Some very special animals can water-ski across the top of the pond, supported by the surface film.

▲ Each tiny duckweed leaf is a single plant. It has no stems and the roots hang down into the water. It floats across the pond carried by the wind. Duckweeds are the smallest plants with flowers.

Water springtails can be very common on a pond, looking like tiny specks of soot. They feed on minute pieces of plant and are so small that their skin is not wetted by the water. They have a tail tucked under their body which can spring open so that they can leap off the water when disturbed.

▼ The water crowfoot is like a buttercup growing in the pond. The floating leaves support the white flowers on their stalks. Leaves under the water have a different shape.

The shining black beetles dashing round in circles are whirligig beetles. They have flattened legs to skate across the surface, but can also swim under the water, and fly off to other ponds. You will see them in small groups in late summer.

The broad-leaved pondweed has leaves underwater and broad green leaves which float on the surface. They lie flat because of a hinge at the end of the leaf stalk.

Waterlilies have larger round leaves and big flowers. They can grow in very deep water with roots sometimes three metres down in the mud.

▼ The water cricket is a small stout pondskater. The adult is colourful, with an orange underside and two orange lines down its back. Crickets and skaters use sturdy front legs to catch insects trapped at the surface. Long legs and hairy bodies spread their weight to stop them falling through the surface film.

▼ This pond skater steers with its back legs and rows with the middle pairs. It hops on the water when disturbed. Mosquito larvae hang beneath the surface taking in air through a tube. Hairs near the mouth filter out tiny plants and pieces of decaying matter. The pupa is also an active swimmer and both pupa and larva will swim away if the surface is touched.

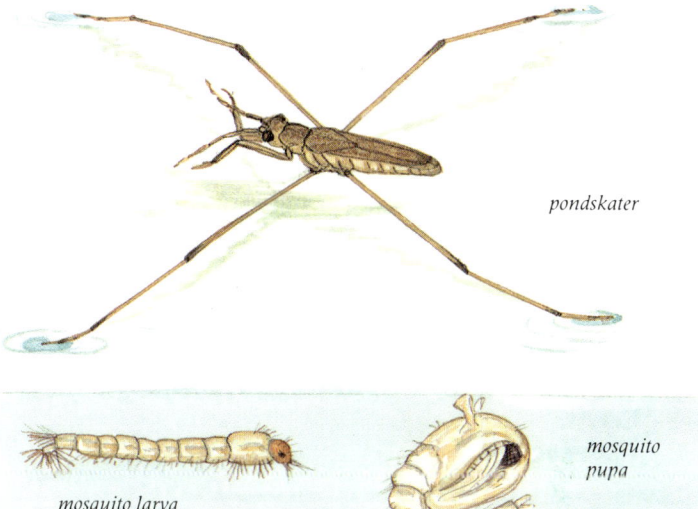

pondskater

mosquito larva

mosquito pupa

Pond facts

Each duckweed leaf produces a bud which becomes a new plant. Duckweeds can very quickly cover an entire small pond! At the end of the summer some leaves sink to the bottom of the pond, well away from the winter ice.

Whirligig beetles have the best of both worlds. With eyes divided into two, they can see along the surface and underwater at the same time.

The Underwater Jungle

Below a pond's surface is a tangled jungle of pondweeds. A number of animals crawl on their stems or hide between their leaves.

▲ Tiny water mites are often brightly coloured and have a rounded body with no separate head. They move so quickly on their hairy legs, it is sometimes difficult to see that they have eight legs. The mites are on long green strands of algae, which are common in stagnant water.

▲ Here a water spider has used its silk to trap an air bubble in an underwater bell-shaped web. It swims out to hunt, taking with it an air supply trapped as a silvery bubble amongst the velvety hairs on its body.

All young mayflies have three long ▶ tails. Mayfly larvae also have leaf-like gills on their bodies to collect oxygen from the water. In this mayfly larva, these gills stick out from the side.

▲ As water plants make food, they also produce oxygen. Bubbles of this important gas can be seen near the leaves of the water starwort. The oxygen will be used by all the pond animals which do not go to the surface to breathe.

Starwort is found in dense green patches. There are pairs of oval leaves on the stem, and at the top a tight rosette of leaves floating just below the surface.

▼ Most pond snails come up to breathe air above the pond. The wandering snail is the most common pond snail. Pond snails feed on algae, rotting pieces of plant and even fish eggs.

The ramshorn snail has a tightly coiled shell and a reddish body. The red colour in the ramshorn's blood helps it to get oxygen from the water.

Canadian pondweed, with its tough leaves and strong stem, does not collapse when taken from the water, like many other water weeds.

wandering snail

ramshorn snail

Canadian pondweed

great pond snail

Pond facts

Each pond snail is both male and female at the same time! One snail can lay up to 500 eggs. If you find snail eggs, you can often see the hatched miniature snails gliding round inside.

Fish, Frogs and Friends

Amphibians can live both on land and in the water where they lay their eggs.

▼ The smooth newt spends much of the year under stones or logs, coming out at night to feed. The smooth newt has olive-brown skin, although the female has a white throat with black spots. In spring, the male has black spots on an orange belly, a crest and striped head. A female newt may lay up to 300 eggs amongst pondweed in the spring and each one hatches into a slim tadpole which hides amongst the plants. They leave the pond at the end of the summer.

◄ The common toad has a dry, warty skin and crawls instead of leaps. Sometimes thousands are squashed while crossing roads. Near some important toad spawning sites, special road signs have been put up to warn motorists.

Toad-spawn is laid as a long row of black eggs. The tadpoles, which are black with a rounded tail, often swim in open water and many are eaten. When they leave the water it is two or three years before they come back as adults.

The familiar common frog lays its eggs in shallow, weedy ponds. The moist skin of the frog is not always the same colour. Often brown or grey, it can be green or yellow. The larger females lay up to 3,000 eggs, called frog-spawn. The tadpoles feed on tiny algae and hide amongst the weeds, but many are eaten by fish, newts and beetles. The feathery gills disappear after four weeks and the back legs have grown by seven weeks. By twelve weeks the tail has gone and they have front legs. Little froglets can be found round a pond in June or July.

frog-spawn

tadpole

▲ In a small pond the stickleback may be the only kind of fish. You can see here how the three-spined stickleback got its name. The brightly-coloured male zigzags around the female and leads her to the nest he has built. She will lay up to 100 eggs. The male looks after the nest and protects the young fish.

Pond facts

Over half of all frogs spawn in garden ponds. Frogspawn has been found in puddles, watering cans, old horse troughs and even in an outside lavatory! In the south, spawn may be found in January and it has been recorded in Cornwall on Christmas Day!

Bugs and Beetles

All bugs which live in ponds have sharp needle-like beaks to pierce plants or animals. When adult, their wings cross down their backs, unlike those of beetles, whose wing cases meet in the middle.

▼ Both this large water stick insect and the water scorpion have a hollow tube to suck in air from above the pond. The front legs are used to grab insects, tadpoles and small fish.

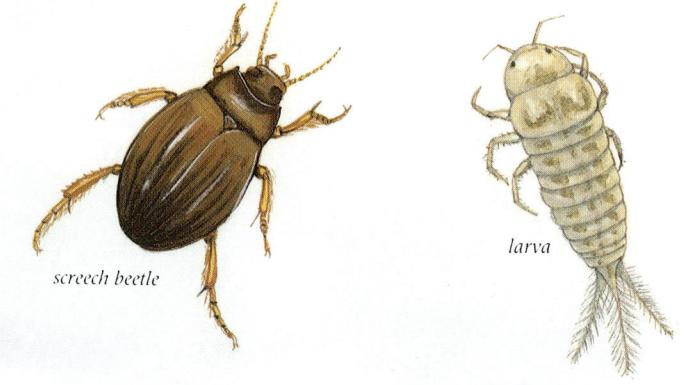

screech beetle

larva

◄ The brightly-coloured screech beetle, with bulging eyes, is not as common as the diving beetle. Its strange larva has three long tails and many gills. Both adults and larvae feed on worms in muddy ponds. If the adult screech beetle is picked up it squeaks! It does this by rubbing its wing cases against the end of its body.

The greater water boatman or ▶
backswimmer is a fierce hunter,
swimming after small animals with its
powerful back legs.

It is easy to recognise the lesser water
boatman even when it is a small larva. It
has a very short beak and swims the
right way up. With hair-fringed legs it
rows through the water, sucking up
algae and tiny pieces of decaying
material.

▼   Beetles also have swimming legs
lined with hairs. There are many types of
diving beetle. All are streamlined and
slippery, and difficult to pick up. The
great diving beetle is one of the largest.
Like the adult, the larva catches quite
large animals, sucking out their juices
through hollow jaws. Hatching from the
pupa at the edge of the pond in spring,
the adult can easily fly to new ponds.

This other common diving beetle is
much smaller. It is red, round and easily
spotted.

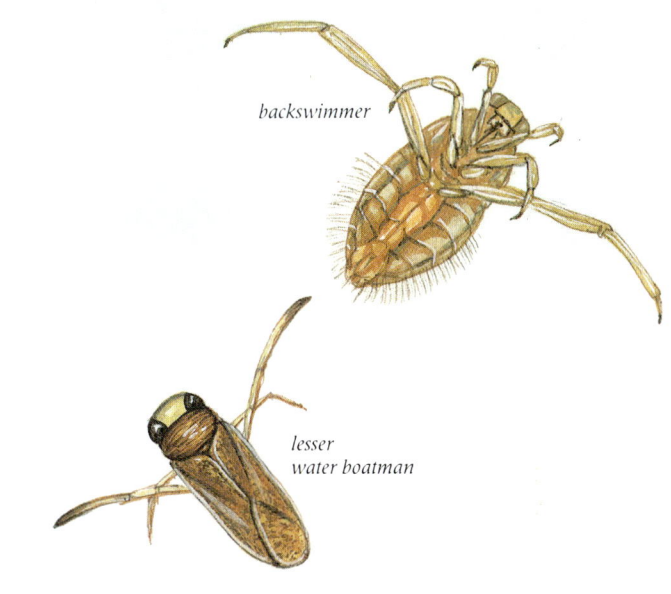

backswimmer

lesser
water boatman

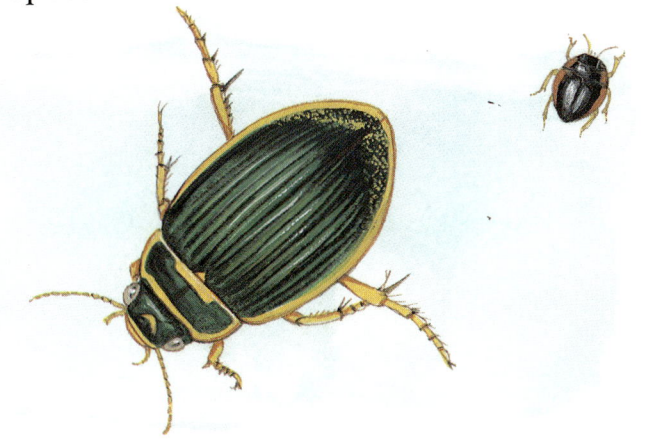

Unlike many adult insects, the great
diving beetle may live for a few years
and visit a number of ponds. It
hibernates during winter, buried in the
mud at the bottom of the pond. If you
quietly approach a pond, you may see
diving beetles and greater water
boatmen hanging from the surface.
Adult beetles take down a bubble of air
under their wing cases. If they stop
swimming, they float back to the
surface. Take care when handling a
greater water boatman. It can bite!

Half-way Down

Water fleas are tiny crustaceans which move in jerks through the water. Using their antennae they appear to jump like fleas.

▼ Water fleas hatch from eggs inside their mother. By using a magnifying lens, these tiny babies can be seen whilst still inside their see-through mother. Water fleas feed by filtering algae and are eaten by all sorts of water animals.

Lying motionless in the water, ▶ completely transparent, is the phantom midge, or ghost larva. You may just see the tiny black eyes and air sacs which help it to float. It grabs small insects and water fleas by the antennae. Stiff hairs at the back form a rudder which helps it to steer.

cyclops

seed shrimp

▲ Other small crustacea swim with the water fleas. Cyclops can move quickly, using antennae and small legs. Females, each with two bags of eggs, are easy to spot.

Seed shrimps look like tiny swimming beans inside their two-sided shells.

▼ The water hog louse, or water slater, looks like the familiar woodlouse found under logs and stones. When it crawls along the muddy bottom it is camouflaged, but it is visible when it climbs on plants such as long strands of algae. Eggs laid by the female can be seen as a white mass under her body. When the young hatch, they are carried under the body of their mother.

▼ The freshwater shrimp is curved and often swims on its side. Other legs are used to hold food, and to move fresh water towards the gills. The larger male is often seen carrying the female, with the eggs or young underneath the mother. Both the shrimp and the water hog louse act as a pond's refuse collectors, feeding on small pieces of decaying plants.

Pond facts

Water fleas can be present in very large numbers. Before winter some eggs in cocoons drop to the bottom instead of developing. These winter eggs are not affected by freezing. Birds may carry them to new ponds. If the pond dries out, they can be carried by the wind.

Life at the Bottom

One of the most unusual pond animals is the caddis fly larva. It hides inside a case of small stones, pieces of leaf or even tiny shells.

This large caddis fly has made its ▶ case from bits of plant cut by its jaws and joined together with sticky silk to form a spiral.

▼ Cranefly larvae wriggling on the pond's bottom will fly from the pond as the familiar daddy-long-legs.

Cockles suck in water through a tube and filter out the food. You may find them partly buried in mud, but they can also climb up the pondweeds. Cockles have been found attached to the feet of wading birds, which carry them to new ponds.

Cockle and two varieties of cranefly larvae

alderfly larva

bloodworm

▼ This short fat larva of the four-spotted chaser dragonfly is a fierce hunter. Here it is seen on gravel, but crawling on the muddy bottom of a pond it is well hidden from other animals. When an insect, tadpole or even a small fish comes close, it pounces with hinged jaws. It can breathe by taking the air from water it squeezes into its bottom. When the dragonfly wants to move fast, it quickly squeezes the water out.

▲ The fully-grown larva of the alderfly is a fiercesome animal. With pincer-like jaws it catches caddis-flies, worms and other animals. It has one long tail and seven pairs of feather-like gills.

When they are disturbed, bloodworms wriggle in the shape of a number eight (8). When adult they are the non-biting midges which dance over the pond in summer like a cloud of smoke.

flatworm

leech

▲ Leeches and flatworms may at first sight look like lumps of dark jelly. Watch carefully and you will see that leeches cling with suckers and move by looping their bodies. They suck the blood of worms, bloodworms and snails. Flatworms glide gracefully and have no suckers.

Pond facts

New garden ponds are important for dragonflies. Many British species have become extinct over the last sixty years and many more are now much rarer. Large dragonflies can take up to five years to develop underwater, shedding their skin ten to fifteen times as they grow.

27

Activities

Pond-watching kit

You can often discover what is happening underwater without catching any animals. Ask an adult to take the top and the bottom from a thick plastic bottle. Cover the edges with tape if they are still sharp. Cover one end with cling film or a clear plastic bag and keep it in place with a thick band.

A net is needed for a closer look. In shallow water use a kitchen sieve or a small aquarium net from a pet shop. To reach out further, use jubilee clips to attach the sieve or net to a cane. By marking the cane every 10cm with a thick waterproof pen, it can also be used as a stick for testing and measuring depth.

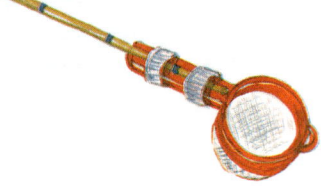

A piedish or an ice-cream or margarine tub is useful to empty your catch into. To sort out the animals use a dish, like an ice-cube tray, which is divided into sections. A magnifier or 'nature viewer' is very useful. Some are big enough for quite large animals and may have a scale on the bottom to help with measuring. A plastic spoon and a fine paint brush are best for transferring your catch.

Don't forget to take this identification book with you, in a plastic bag, wellies and a towel to dry your hands if it is cold. When visiting a pond remember the pond-watching safety code on page 7.

Pond dipping

Remember to approach the pond quietly and look carefully before you start to dip. Put clear water into your trays and viewer. Sweep your net quickly and turn it into the tub. After removing stones and pieces of plant, look carefully and sort out your catch. Don't sweep too much in one area and remember to put everything back when you have finished.

Try sweeping across the surface, in open water and amongst water weeds. Take a sieve full of mud from the bottom and wash it backwards and forwards in the water. Fine clay will fall through leaving the animals.

Investigating pond animals

You now know where animals are living, but how do they live? Put a collection of animals into a deep clear jar of pond water, for example a large lemonade bottle with the top cut off. How do they breathe? Which animals come up to the surface for oxygen? Which animals have gills to let in more oxygen? Can you see the body or the gills moving to get fresh water? If the animal swims, what special features does it have to help it?

Look closely to see how water crickets and pond skaters seem to be walking on a skin. Back home, fill a tub with water. Place a needle on a piece of blotting paper and put it onto the water. The paper will slowly sink. Can you make the needle float, held by the skin, or surface tension?

You may see some animals feeding. Carnivores often have sharp jaws. Put a

dragonfly or alderfly larva in a tub of water and slowly move the tip of a pencil towards its head. What happens?

It is exciting to study how frogs develop from the eggs or spawn, but only take spawn from a garden or school pond which has plenty.

Take a small amount of spawn, about half a cupful. Put it into a tank of water that has been left standing for a few days. You could put in a few water plants. Feed the spawn every few days with bits of boiled lettuce leaves or with a few pellets sold to feed rabbits. Change the water when it gets murky. As the tadpoles develop, make sure they have something to climb up, and when front legs grow, tip the tank so that it is shallower at one end. Release the froglets into damp long grass, but never back into a pond or they could drown.

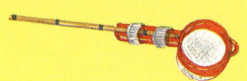

Diary

If you visit a pond regularly, keep a diary. How do your observations compare with the *Year in the life of a pond* in the Introduction? When are the plants in flower or seed? When do you see the first frog spawn, the first dragonfly, and the young of water birds? Which birds visit to feed and drink?

Pond conservation

Small ponds quickly fill up with large plants and mud and may even dry out in the summer. Neglected, they may also fill up with litter. Why not adopt a pond? But first, if it is not your own, or the your school's, make sure you have permission.

You may need to top up the water level in summer. If you have to use tap water, leave it standing for a few days and only put in a few buckets at a time.

In early autumn, remove a few bucketfuls of sludge and any dead leaves. You may also have to pull out some of the taller plants and water-weed to make sure that there is some open water for next year.

In winter, try to keep some water ice-free by floating a ball in it. Sweep off any snow round the edges to let in light.

Magnetic food-web game

Trace the outline of each of these pictures on to card. Bloodworm × 2 means that you need to make two bloodworm cards. Colour all your cards and on the back write on the animal's name and how it feeds. In this pond the stickleback will eat many of the smaller animals.

Make a deep pond from a cardboard box. You could paint it and decorate it with pictures of pond plants. Make each player a fishing rod using a small stick, to which is attached a piece of string and tied to a small magnet. Attach paper clips to each card and put the cards into the pond.

Imagine you are a kingfisher. To make sure that there is a nice plump stickleback to eat you need to complete the food chain. Collect eight plants, four herbivores, two carnivores, and one stickleback. Each player takes it in turn, putting back cards they do not need. To make the game more difficult, collect the cards in order.

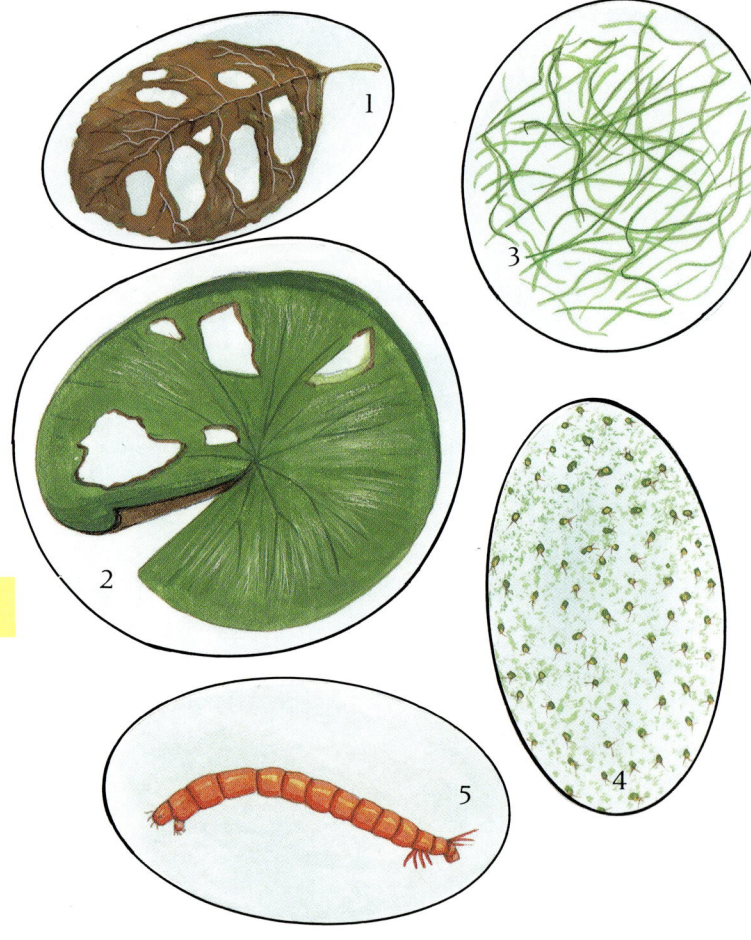

Since a stickleback can eat herbivores and carnivores you could catch four herbivores and two carnivores or six herbivores. This would stop the other player from completing a food chain!

1 Dead leaf PLANT ×4
2 Waterlily leaf PLANT ×4
3 Hair-like algae PLANT ×4
4 Microscopic algae PLANT ×4
5 Bloodworm HERB ×2
6 Water flea HERB ×2
7 Wandering snail HERB ×2
8 Lesser water boatman HERB ×2
9 Diving beetle CARN ×2
10 Damselfly larva CARN ×2
11 Stickleback CARN ×2

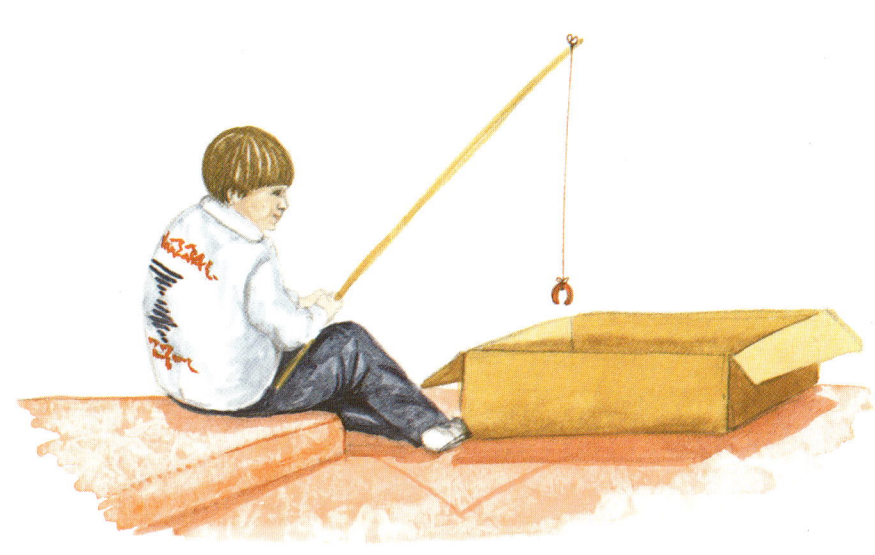

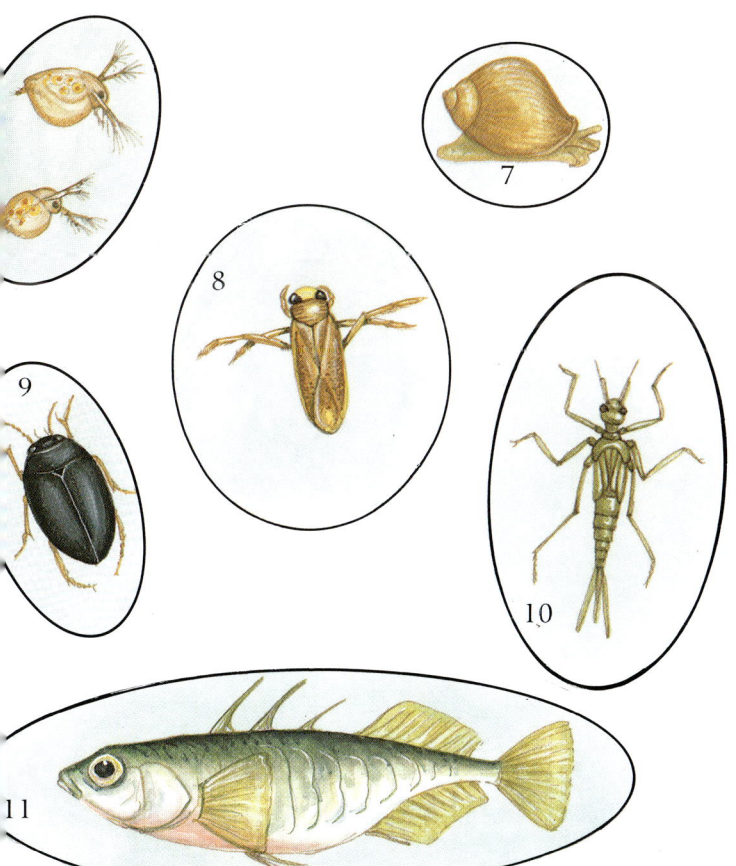

If you have enjoyed reading this book, why not take part in a pond survey? Pondwatch surveys have been organised by the Wildfowl and Wetlands Trust in association with WATCH, the wildlife and environment club. WATCH has also run projects on streams, dragonflies and frogs. There are WATCH groups in your area run by your local Wildlife Trust. For more information write to

The Wildfowl and Wetlands Trust
Slimbridge
Gloucestershire
GL2 7BT

WATCH
The Royal Society for Nature Conservation
Witham Park
Waterside South
Lincoln LN5 7JN

Index